I0606283

Early
ANIMAL
Encyclopedias

# CATS

by Kathy MacMillan

Early Encyclopedias

An Imprint of Abdo Reference
abdobooks.com

**abdobooks.com**

Published by Abdo Reference, a division of ABDO, PO Box 398166, Minneapolis, Minnesota 55439.


Printed in China
102022
012023

Editor: Marie Pearson
Series Designers: Candice Keimig, Joshua Olson

Library of Congress Control Number: 2022940652

Publisher's Cataloging-in-Publication Data

Names: MacMillan, Kathy, author.
Title: Cats / by Kathy MacMillan
Description: Minneapolis, Minnesota: Abdo Publishing, 2023 | Series: Early animal encyclopedias | Includes online resources and index.
Identifiers: ISBN 9781098290399 (lib. bdg.) | ISBN 9781098275716 (ebook)
Subjects: LCSH: Cats--Juvenile literature. | Cats--Behavior--Juvenile literature. | Zoology--Juvenile literature. | Encyclopedias and dictionaries--Juvenile literature.
Classification: DDC 636.8003--dc23

# CONTENTS

## INTRODUCTION

People have kept cats to control rodents for thousands of years.

# The History of Cats

A long time ago, all cats were wild. About 10,000 years ago, people started growing and storing grain. Mice and rats ate the grain. Wild cats came to eat the mice and rats. People liked this. So they gave the cats shelter.

Over the years, these cats let people get close. They even let people pet them. Ancient Egyptians kept cats in their homes as pets. Ancient Greeks and Romans also kept cats. Cats helped control pests on ships and farms.

## Cat Breeds

In the 1800s, cat shows and cat clubs became popular. People started breeding cats to look or act a certain way. Today there are more than 70 different cat breeds. These cats are much smaller than the biggest wild cats. Most are 20 to 28 inches (51 to 71 cm) from head to rump. But they still have long, strong bodies and claws they can draw in, just like their wild cousins have.

## ABYSSINIAN

Abyssinians are loyal.

# Appearance

Abyssinians are lean but muscular. They have long legs and large ears. Their coats can be cinnamon, blue, fawn, or reddish brown. Their coats are ticked. These cats often have lighter fur on their bellies. The tips of their tails usually have darker fur.

**Weight:**
8 to 12 pounds
(3.6 to 5 kg)

## Behavior

Abyssinians have a lot of energy. They are smart and interested in everything. They love to play, jump, chase, and climb. These cats enjoy spending time with their people.

## History

The first Abyssinian cat was shown in an 1871 cat show. The owner said the cat came from Abyssinia. That was the name for the African country now called Ethiopia. This is how the breed got its name. But scientists think the breed may have started in India.

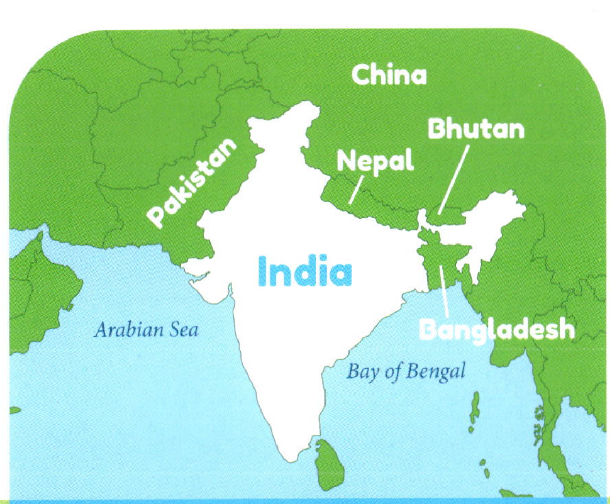


**From:** India

# AMERICAN BOBTAIL

Some American bobtails play fetch.

## Appearance

American bobtails have long, strong bodies and large paws. They have short, stumpy tails like wild bobcats. Their fur can be short or long. The coats come in many colors and patterns. The tips of these cats' ears sometimes have tufts of fur.

## Behavior

American bobtails are friendly and smart. They love being with people and playing games. These cats can learn to walk on a leash.

## History

In the 1960s, a couple found a cat with a short tail in Arizona. They took him home. He had kittens with the couple's long-tailed cat. The kittens had short tails. These cats became the American bobtail breed.

**Weight:**
5 to 10 pounds (2.3 to 5 kg)

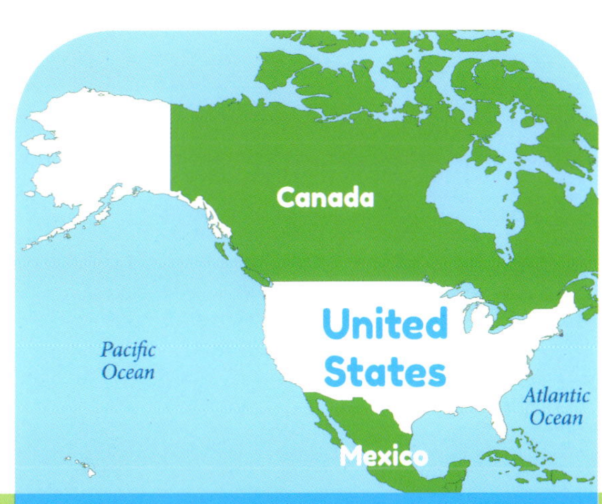


**From:** The United States of America

# AMERICAN CURL

American curls have big, wide eyes.

## Appearance

American curl kittens are born with straight ears. Their ears start to curl backward a few days later. They can have short or long hair. Their fur comes in many different colors and patterns.

**Weight:**
7 to 11 pounds (3.2 to 5 kg)

## Behavior

American curls are friendly and smart. They like to play. Some even like to play in water. These cats often follow their owners around the home. They need lots of attention.

## History

In 1981, a California couple found a stray kitten with curled ears. They brought her home. Later, her kittens were born with curled ears. Breeders liked how the cats looked, so they bred more.

**From:** The United States of America

## Appearance

American shorthairs have powerful bodies. Their fur is short and thick. Their coats may be any color or pattern. These cats have round, flat faces.

## Behavior

American shorthairs are good at hunting mice and insects. They are friendly and gentle. These cats are good pets for many homes.

## History

In the 1600s, cats came to North America with settlers from Europe. The cats hunted mice on the ships. Then the cats spread to farms all over North America. People recognized them as a single breed. In the 1960s, this breed was named the American shorthair. It is now one of the most popular breeds.

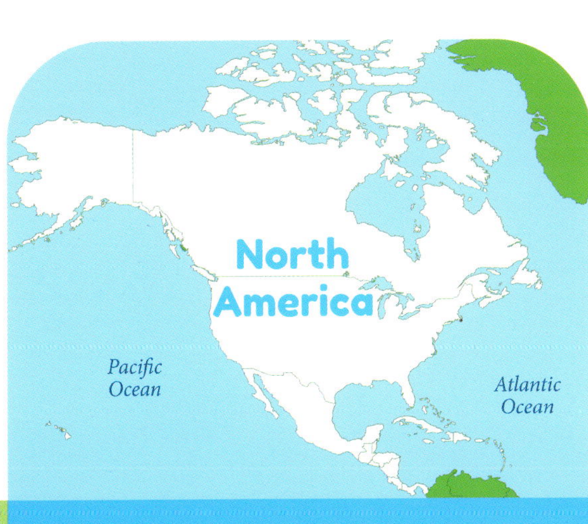


**Weight:**
6 to 15 pounds (2.7 to 7 kg)

**From:** North America

# AMERICAN WIREHAIR

American wirehairs purr a lot.

## Appearance

The American wirehair's fur is bent or crimped. The coat looks and feels coarse. It may be any color. American wirehairs also have curly whiskers. Their bodies are rounded. They have short noses.

**Weight:**
8 to 15 pounds (3.6 to 7 kg)

## Behavior

American wirehairs are shy at first. But once they get to know someone, they are friendly. Wirehairs like to play and watch birds. They hunt insects. These cats tend to be quiet.

## History

The first American wirehair was born on a farm in New York in 1966. His parents were shorthair cats. But this kitten had wiry hair. A cat breeder saw him. She liked his coat. She took him home. She bred more kittens with wiry coats.

**From:** The United States of America

# AUSTRALIAN MIST

The Australian mist's eyes are green.

## Appearance

Australian mist cats have short coats. Their coats can be spotted or marbled. Some of their fur is ticked. They look as if they are covered in mist.

## Behavior

Australian mists were bred to be indoor cats. They like to run and play fetch. They also love to cuddle. These cats usually get along with other pets.

## History

This is the only cat breed from Australia. In 1976, a breeder crossed Burmese, Abyssinian, and other shorthaired cats. The new breed had spots. The breeder called it the spotted mist. Later, cats with marbled coats became part of the breed too. The breed's name changed to Australian mist in 1998.

**Weight:**
8 to 13 pounds (3.6 to 6 kg)

**From:** Australia

# BALINESE

Balinese have blue eyes.

## Appearance

Balinese have thin bodies. Their heads are long. They also have long, silky hair. The fur on their tails may be up to 5 inches (13 cm) long. Their fur comes in many colors. No matter what the color is, the fur is pointed.

**Weight:**
6 to 11 pounds (2.7 to 5 kg)

## Behavior

These cats are curious. They love attention. They are noisy and meow a lot. They like to climb on people's shoulders.

## History

Sometimes Siamese cats would have longhaired kittens. In the 1950s, two Americans decided to breed these kittens. They chose the name Balinese for this new breed. The graceful look of the cats made them think of dancers from Bali, an island in Indonesia.

**From:** The United States of America

## Appearance

Bengals have spots. This makes them look like leopards, jaguars, or cheetahs. Bengals have short, silky fur. Some even look like they glitter. Their eyes are large.

**Weight:**
Usually 8 to 15 pounds (3.6 to 7 kg), but sometimes up to 22 pounds (10 kg)

## Behavior

Bengals like action. They want to be in the middle of things.

They enjoy climbing. They perch up high. Unlike most cats, Bengals love water. A Bengal might even jump into the shower with its owner.

## History

In 1963, a breeder crossed the small Asian leopard cat with house cats. The kittens were gentle like house cats. But they looked like wild cats. These new cats became the Bengal breed.

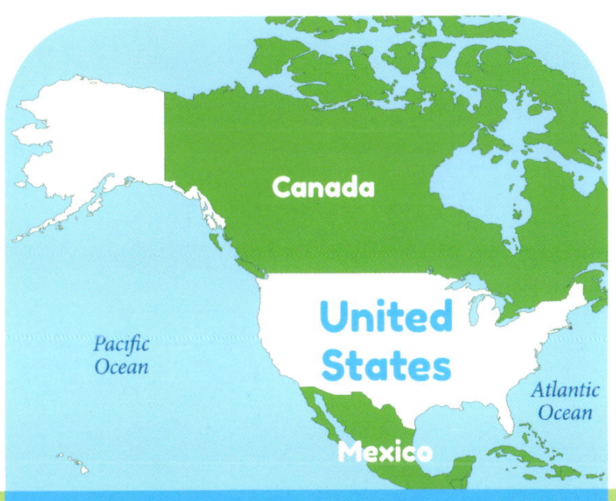


**From:** The United States of America

## Appearance

Birmans are stocky. They have long, fluffy fur. They are born all white. Later, they become pointed. These points can be many colors. Birmans have white paws.

## Behavior

Birmans like a peaceful life. They are usually quiet. But they do not like to be alone.

They want to be around people. They get along well with other pets.

## History

No one knows where the Birman came from. Stories say that Birmans were once kept by priests in Burma. This is a country in Southeast Asia. Some Birmans were shown in a cat show in France in the 1920s. They were called Birmans after the French name for Burma.

**From:** Unknown

**Weight:**
7 to 14 pounds (3.2 to 6 kg)

## BOMBAY

Bombays look like little panthers.

## Appearance

Bombays have shiny, short, black coats. Their noses and paw pads are black too. They have gold or copper eyes. These cats have round heads with short noses.

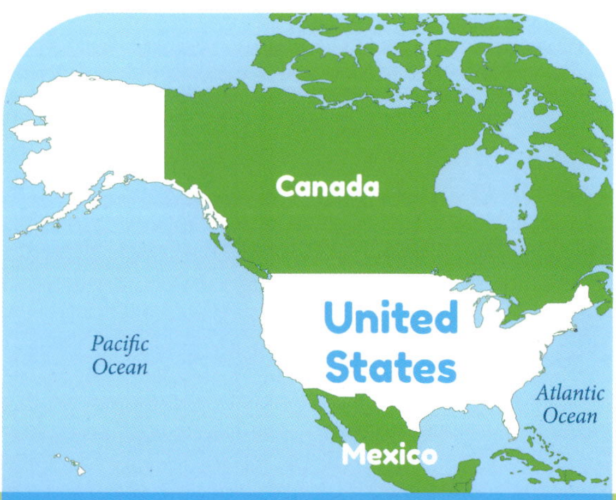


**From:** The United States of America

## Behavior

Bombays are friendly and sweet. They might greet strangers at the door. Bombays can be silly too. They do not like to be alone. They like to play games. But they also sit in people's laps.

**Weight:**
6 to 11 pounds (2.7 to 5 kg)

## History

A cat breeder in Kentucky created Bombays in 1958. She wanted a cat that looked like a black panther from India. She bred a Burmese with an American shorthair. Bombay is the English name for Mumbai, a city in India.

BRITISH LONGHAIR

The best known British longhair color is gray.

## Appearance

British longhairs are fluffy. They have long, thick fur. They have round faces and eyes. These cats come in many different colors. Their round faces make them look like they are smiling.

## Behavior

These cats can be shy. But when they get to know a person, they are friendly. They like to play, but they do not mind being alone. They are calm cats.

## History

Between 1914 and 1918, breeders crossed British shorthairs with Persians. Later, other shorthaired cats were added. The new breed had long hair like a Persian. But its face was less flat and more rounded like a British shorthair. People called this new breed the British longhair.

**From:** The United Kingdom

**Weight:**
8 to 16 pounds (3.6 to 7 kg)

## BRITISH SHORTHAIR

British shorthairs have thick fur.

# Appearance

British shorthairs have round faces and eyes. Their round faces make them look like they are smiling.

**From:** The United Kingdom

The most popular coat color is gray. But these cats can be many other colors too.

**Weight:**
8 to 16 pounds (3.6 to 7 kg)

## Behavior

These cats are calm and quiet. They like to play, but they do not get wild. They do not mind being alone sometimes. But they also like to be with people.

## History

British shorthairs are one of the oldest cat breeds in the world. Some experts think the cats that later formed this breed came to what is now the United Kingdom with the Romans in 55 BCE. These cats would kill mice and other pests on farms.

## BURMESE

Burmese have powerful bodies.

## Appearance

Burmese cats have round heads. They have wide-set, yellow or gold eyes. Their fur is glossy and short. Their coats may be sable, champagne, blue, or platinum. Their bellies are a lighter color.

## Behavior

These cats are playful and friendly. They love people. They will follow their owners around.

They do not like to be alone. Sometimes they can be bossy.

## History

In 1930, a sailor got a brown cat named Wong Mau from Burma. He gave Wong Mau to a doctor in California. The doctor bred her with a Siamese cat. Some of her kittens were dark brown. They were the first Burmese cats.

**From:** The United States of America

**Weight:**
6 to 14 pounds (2.7 to 6 kg)

# BURMILLA

The Burmilla has a strong and beautiful body.

## Appearance

Burmillas have sparkling, silver coats. Their fur may be shaded. The fur's tips may be other colors such as black, brown, or lilac. It may be short or long. Their green eyes are set far apart.

## Behavior

Burmillas love to play. But they are also quiet and gentle. They like people and children.

**From:** The United Kingdom

They get along with other pets. If bored, Burmillas will find their own fun.

## History

In 1981, a chinchilla Persian had kittens with a Burmese. Chinchilla Persians are white with black-tipped fur. The kittens had short, thick coats that were black shaded. The name Burmilla is a cross between Burmese and chinchilla.

**Weight:**
8 to 12 pounds (3.6 to 5 kg)

# CHARTREUX

Chartreux have woolly fur.

## Appearance

Chartreux (shahr-TROO) have short, blue-gray fur. The tips of their fur may be silver. They have orange eyes. Their ears are set high on their heads. This makes them look alert.

**From:** Syria, possibly

## Behavior

These cats are calm and quiet. But they may chirp to get someone's attention. They like to be with people. They enjoy sitting on laps. Some like to play fetch.

## History

The Chartreux is one of the oldest cat breeds. People do not know for sure where the breed started. Chartreux may have come from Syria. But they may have gotten their name from a fine Spanish wool called Chartreux. People may have thought the cat's fur felt like the wool.

**Weight:**
12 to 16 pounds (5 to 7 kg)

# CHAUSIE

Chausies have large ears.

## Appearance

The chausie (CHOW-see) is a large, long cat. It can be up to 22 inches (56 cm) long from nose to rump. These cats have short fur. Their fur can be brown ticked tabby, black grizzled tabby, or solid black. Their tails are shorter than those of many other breeds.

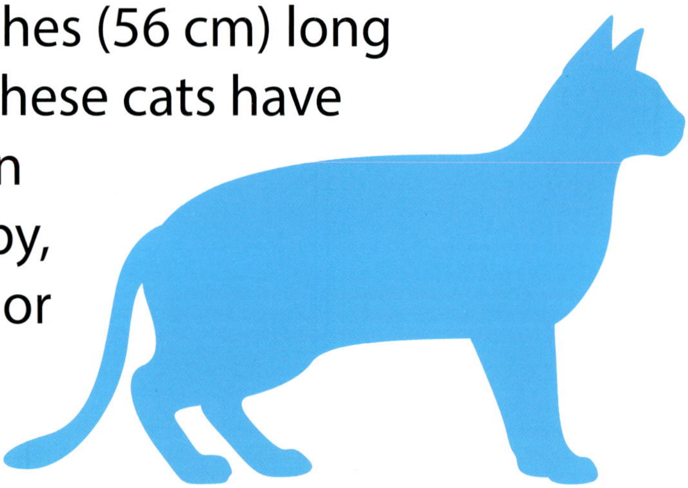

**Weight:**
15 to 25 pounds (7 to 11 kg)

## Behavior

These cats love to move. They like people and do not want to be alone. Chausies are active. They may get into trouble. They enjoy learning tricks. They also like to play in water.

## History

The first known crossing of a domestic cat with a jungle cat was in 1960. Jungle cats are small wild cats from northern Africa, the Middle East, and Asia. Breeders in the United States decided to make this cross into a breed, the chausie.

**From:** The United States of America

# CORNISH REX

The Cornish rex has a curvy body.

## Appearance

Cornish rexes have short, soft, curly coats. They have soft undercoats and no top coats. Their fur comes in many colors and patterns. These cats have large ears. They have long legs and tails.

## Behavior

Cornish rexes like to move and play. They run and jump.

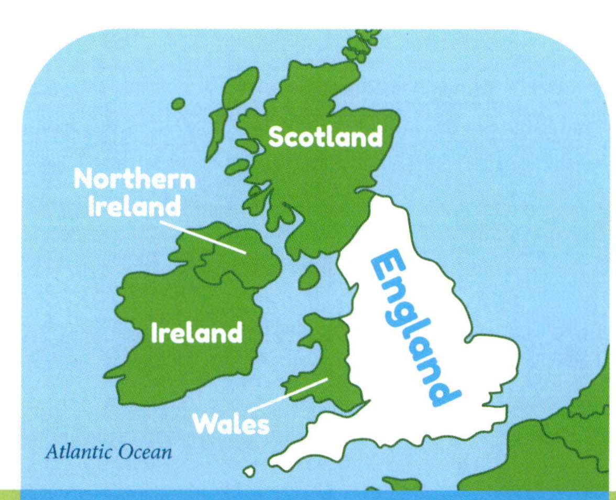


**From:** England

They enjoy playing fetch. Sometimes, they even throw a toy to play fetch alone. They love to cuddle and be with people. They eat a lot.

## History

The first Cornish rex was born in Cornwall, England. In 1950, a farm cat had a litter. One of the kittens had curly fur. People wanted these cats. So breeders bred more Cornish rexes.

**Weight:**
6 to 10 pounds (2.7 to 5 kg)

# CYMRIC

Cymrics may have no tail or they may have a stump tail.

## Appearance

The Cymric (KUM-rick) has long, silky hair. Its body is short and round. Its fur can be any color or pattern. The Cymric's hind legs are longer than its front legs.

## Behavior

These cats are calm and loving. They will trill to show how they feel. They can be silly. They like to play and jump.

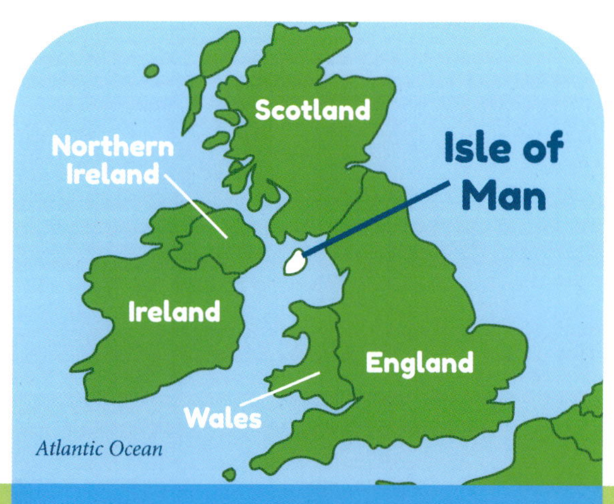


**From:** Isle of Man

## History

Cymrics came from the Isle of Man in the Irish Sea. Some cats there were born without tails. They were called Manx cats. Viking ships brought longhaired cats to the island. They had kittens with the Manx cats. Some had longer fur. Today, some people say Cymrics are longhaired Manx. Others say they are a separate breed.

**Weight:**
8 to 12 pounds (3.6 to 5 kg)

DEVON REX

Devon rexes that live together will often snuggle together.

## Appearance

Devon rexes have big eyes. Their ears are large. These cats have thin, soft, curly fur. The Devon rex has an undercoat without a top coat. The coat may be any color or pattern. Devon rexes have long faces. They also have long legs and tails.

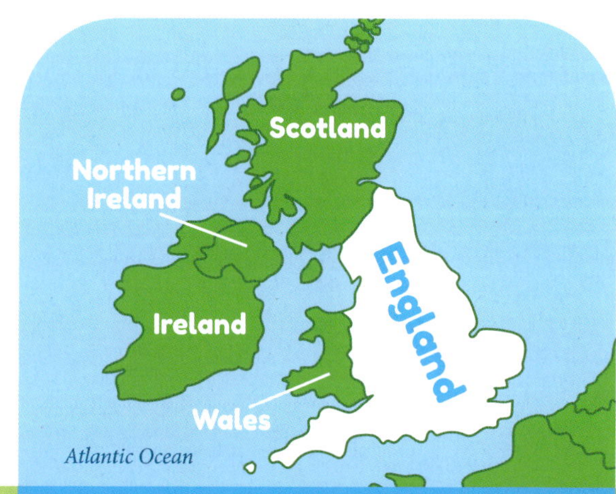


**From:** England

## Behavior

These cats love people. They like to cuddle. They also love to play. They jump and explore. They love food too.

## History

In 1960, a farm cat in Devon, England, had kittens. One of them had a curly coat. He was the first Devon rex. Later, his owner bred more cats like him.

**Weight:**
6 to 9 pounds (2.7 to 4.1 kg)

# Appearance

Some Don sphynx are born hairless. Others are born with fur. As they get older, the fur falls out. Some keep a thin layer of fur. But most are hairless. They have wrinkles on their faces, necks, legs, and bellies. Don sphynx have large, wide ears. They have webbed toes.

Don sphynx's eyes slant up.

# Behavior

Don sphynx are loving, gentle cats. They are also active and friendly. They are very smart. Sometimes they make up their own games.

## History

In 1987, a woman in Russia rescued a stray kitten. As the kitten grew up, her hair fell out. When she had kittens, they lost their hair too. This started the Don sphynx breed.

**From:** Russia

**Weight:**
6 to 12 pounds (2.7 to 5 kg)

# EGYPTIAN MAU

## Appearance

Egyptian maus have spotted coats. They have long hind legs. Egyptian maus can have silver, bronze, or smoke coloring. All Egyptian maus have large, green eyes. Their legs and tails have bands of color.

## Behavior

These cats like to be in control. They will cuddle only if it is their idea.

**From:** Egypt

They do not like loud noises. They can run as fast as 30 miles per hour (48 kmh). They like water.

## History

The Egyptian mau comes from Egypt. The breed came from spotted wild cats who became house cats. *Mau* means "cat" in Egyptian. Ancient Egyptian art showed Egyptian maus.

**Egyptian maus like climbing and jumping.**

**Weight:**
6 to 14 pounds (2.7 to 6 kg)

## Appearance

Exotics have round heads and bodies. They have flat faces and fat cheeks. Their eyes and ears are set far apart. They have short legs and tails. Their thick fur can be any color or pattern. It can be long or short.

**From:** The United States of America

## Behavior

These cats like to play sometimes. But they also like to rest in people's laps. They do not make a lot of noise. They may chirp or meow quietly.

**Weight:** 6 to 13 pounds (2.7 to 6 kg)

## History

In the 1950s, breeders crossed Persians with the American shorthairs and other breeds. The new cats had bodies and faces like Persians. They were called exotics.

**Exotics are good with children and pets.**

# HAVANA BROWN

Havana browns love food.

## Appearance

Havana browns have red-brown fur. It is short and glossy. These cats have wide-set, green eyes. Their large ears tilt forward. Their whiskers are brown.

**Weight:**
6 to 10 pounds (2.7 to 5 kg)

## Behavior

These cats are playful and curious. They touch and hold things with their paws.

Havana browns like to be close to people. They are not noisy cats. They chirp and trill softly.

## History

In the 1950s, people in England bred Siamese with other cats. They called the new breed Havana. A breeder in the United States

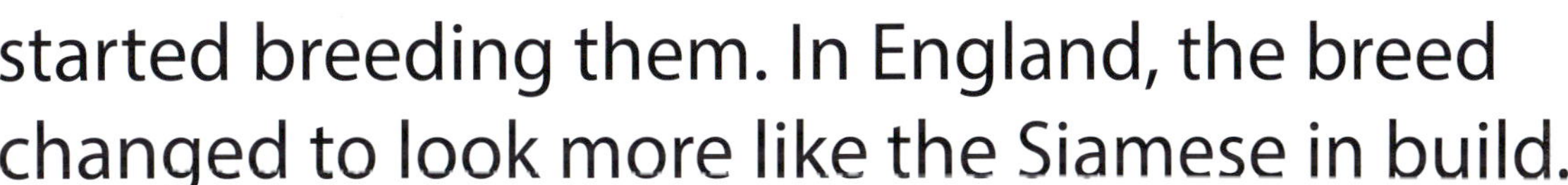

started breeding them. In England, the breed changed to look more like the Siamese in build. The American breed still looked like the original cats bred in England. So these cats got a different name, Havana brown.

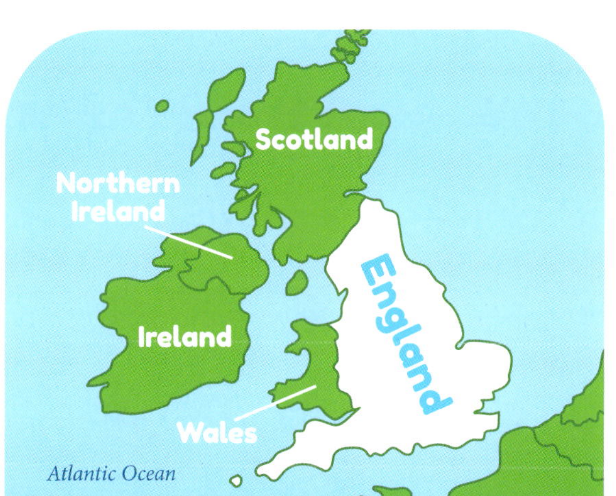



**From:** England

## HIMALAYAN

Himalayans have blue eyes.

## Appearance

Himalayans have long, silky fur. They have round heads and bodies. Their eyes and ears are set far apart. They have short noses. Their fur can be any color. But it is always pointed.

**Weight:**
7 to 12 pounds (3.2 to 5 kg)

## Behavior

These cats are very smart. They tune in to their owners' feelings. They are sweet and like to cuddle. They also like to play. They love to bat things with their paws.

## History

In the 1950s, breeders crossed Persians with Siamese. They wanted a cat with the long, silky hair of a Persian. But they wanted it to have the color pointed pattern of a Siamese. This created the Himalayan.



**From:** The United States of America

# JAPANESE BOBTAIL

The Japanese bobtail's back legs are longer than its front legs.

## Appearance

Japanese bobtails have short, fluffy tails. Some have tails that are bent or curled. Their fur can be long or short. It comes in many colors and patterns. These cats are small. They are only about 12 inches (30 cm) long from nose to rump.

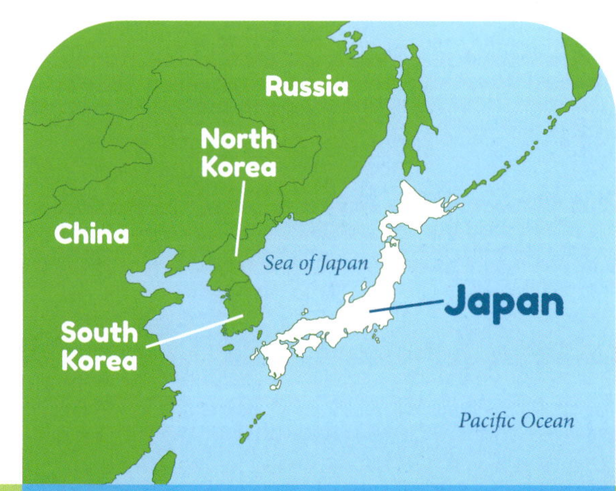


**From:** Japan

## Behavior

These cats do not like to sit still. They play and explore. They carry things in their mouths. They have soft voices. They may even sound like they are singing.

## History

Buddhist monks brought these cats to Japan from China or Korea more than 1,000 years ago. Japanese people say these cats are good luck. Many stores have a waving Japanese bobtail figure in the doorway.

**Weight:**
5 to 10 pounds (2.3 to 5 kg)

# KHAO MANEE

## Appearance

Khao manee (KOW MAH-nee) cats have shining, white coats. Their fur is short and smooth. One eye can be a different color than the other. Khao manees have strong, lean bodies. Their heads are shaped like hearts.

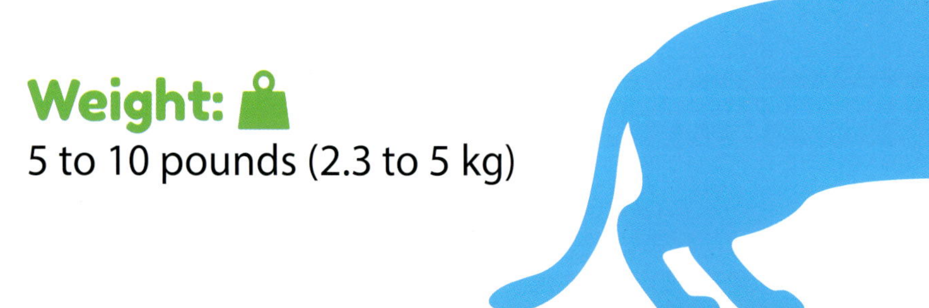

**Weight:**
5 to 10 pounds (2.3 to 5 kg)

## Behavior

These cats are playful and smart. They will greet strangers at the door. They do not like to be alone. Sometimes they get into trouble.

## History

These cats are from Thailand. They are shown in a book from more than 700 years ago. These cats were kept in palaces. People thought they were good luck. Khao manees with two different-colored eyes were thought to be even luckier.

**The khao manee's nickname is the "white gem."**

**From:** Thailand

Many Korats enjoy food puzzle toys.

## Appearance

The Korat (koh-RAHT) has short, silver-blue fur. The fur seems to shine. Each hair is ticked with silver, then blue gray, then silver. These cats have large, green eyes. Their heads are shaped like hearts. Their large ears tilt forward.

## Behavior

Korats love their people. They will sit very close to them. Korats like to cuddle.

**From:** Thailand

They do not like to be alone. They love to play. But they are also gentle. They dislike loud noises.

## History

These cats come from an area in Thailand called Korat. An old book shows Korat cats. It was written between 200 and 600 years ago. People in Thailand say that Korats are good luck.

**Weight:**
6 to 10 pounds (2.7 to 5 kg)

# KURILIAN BOBTAIL

Kurilian bobtails are good hunters.

## Appearance

Kurilian bobtails have short tails like wild bobcats. Their tails may be bent or curled. They have strong bodies. Their fur may be short or long. It comes in many different colors and patterns. These cats can have tufts on their ears.

## Behavior

These cats are gentle and friendly. They love to be petted. They enjoy being with people.

They are good jumpers and like to climb high. They will hunt mice and catch fish.

## History

These cats come from the Kuril Islands. This chain of islands is part of Russia. Kurilian bobtails have lived there for more than 200 years.

**Weight:**
8 to 15 pounds (3.6 to 7 kg)

**From:** Kuril Islands

# LAPERM

## Appearance

LaPerm kittens are born with wavy or straight hair. Then it falls out. It grows back curly. It can be short or long. It can be any color or pattern. LaPerms have long, curly whiskers.

## Behavior

These cats love to be with people. They may pat a person's face with their paws.

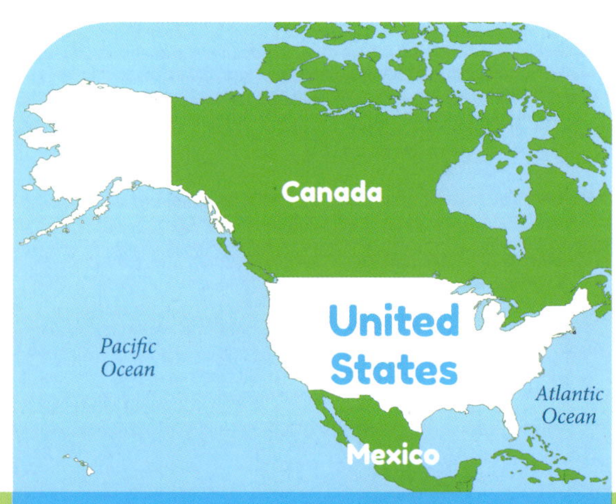


**From:** The United States of America

They like to curl up in people's laps. They are gentle. But they have lots of energy.

**Weight:**
6 to 10 pounds (2.7 to 5 kg)

## History

In 1982, a barn cat in Oregon had kittens. One of the kittens had no fur. Later, she grew soft, curly fur. When she grew up, she had kittens with curly fur too. The owner called the breed LaPerm.

**A LaPerm's fur feels like lamb's wool.**

The lykoi's name comes from the Greek word for wolf.

## Appearance

Lykoi (LIE-koy) can be any color or pattern. These cats have no undercoat. Their fur looks rough. But it feels soft. They have no hair around their noses or eyes. Their legs and feet do not have a lot of fur.

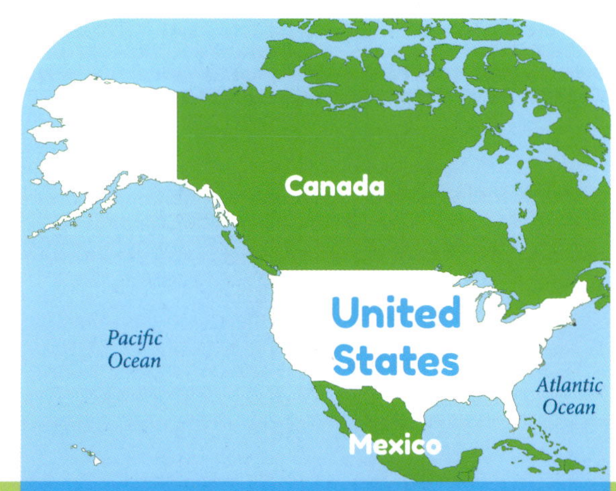


**From:** The United States of America

## Behavior

These cats enjoy being around people. They like warm laps. But they like playing better. They hunt, chase, and pounce.

## History

The first lykoi were found in a Virginia shelter in 2010. People wondered if they were sick. A doctor said the cats were healthy. They just had a strange fur pattern. People liked how the cats looked. So they bred more.

**Weight:**
6 to 10 pounds
(2.7 to 5 kg)

MAINE COON CAT

## Appearance

Maine coon cats are large. They may be up to 30 inches (76 cm) long from nose to rump. Their long fur may be any color or pattern. It is a little oily. This keeps water off. Their large paws and ears are tufted.

## Behavior

These cats follow people around. But they do not like to sit in people's laps.

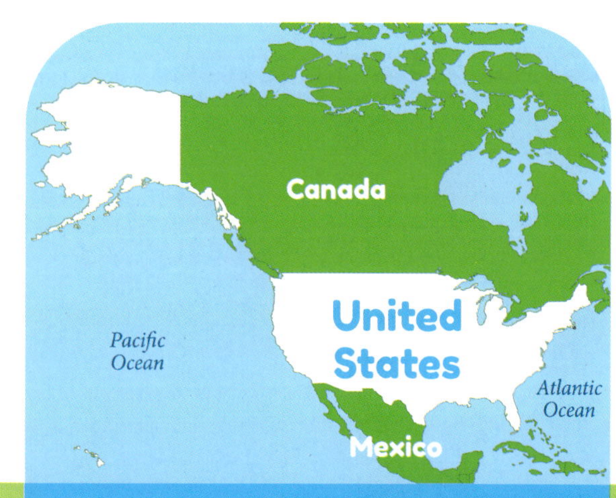


**From:** The United States of America

They are playful even when they get older. They have soft voices. They chirp or trill.

## History

These cats come from Maine. People thought they looked like raccoons. That is how they got their name. Farmers showed these cats at fairs in the 1860s.

**Weight:**
12 to 22 pounds (5 to 10 kg)

**Maine coon cats love water.**

# MANX

Some Manx enjoy carrying their toys around.

## Appearance

Many Manx are tailless. But kittens with and without tails can be born together. These cats' front legs are shorter than their back legs. Their heads, bodies, and cheeks are round. Their short, glossy fur may be any color or pattern.

**Weight:**
8 to 12 pounds (3.6 to 5 kg)

## Behavior

These cats are good runners. They can turn quickly. Their strong back legs help them jump high. They play by hunting their toys.

## History

These cats come from the Isle of Man in the United Kingdom. They have been around for more than 200 years. British shorthairs there may have had some tailless kittens. Soon more cats without tails were born.

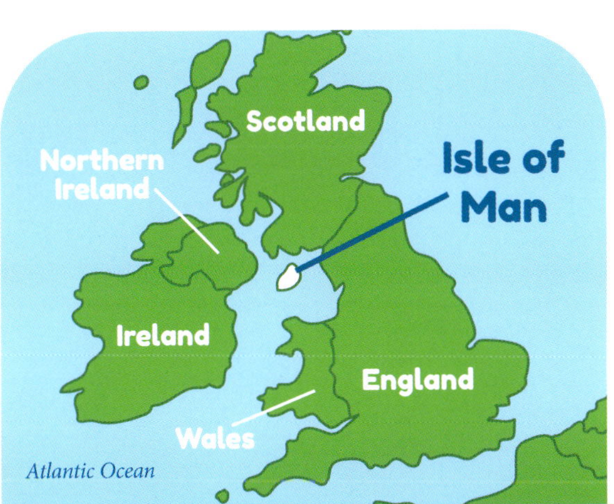


**From:** Isle of Man

# MINUET

## Appearance

Minuets have short legs. Their heads and bodies are round. Their big eyes are round too. Their thick fur may be short or long. It can be any color or pattern.

**Minuets have long tails.**

## Behavior

These cats are gentle. They love people and like to cuddle. But they are also curious. They enjoy playing.

**From:** The United States of America

They climb and jump. Their legs are short, but they can still run.

**Weight:**
7 to 8 pounds (3.2 to 3.6 kg)

## History

A breeder crossed Persians and Munchkins in the 1990s. The kittens had short legs like Munchkins. But they had thick fur and were gentle like Persians. The new breed was eventually named minuet. This comes from a French word meaning "small."

MUNCHKIN

Munchkins enjoy snuggling.

## Appearance

Munchkins have very short legs. Their tails are as long as their bodies. Their thick fur may be short or long. It comes in many colors and patterns.

**From:** The United States of America

## Behavior

These cats are playful. They can run fast. They are smart and curious. They will stand up on their hind legs to see better. Munchkins may steal and hide shiny things. They love people.

**Weight:**
5 to 10 pounds (2.3 to 5 kg)

## History

In 1983, a breeder found a cat with short legs. She bred more cats with short legs. The breed's name comes from the book *The Wonderful Wizard of Oz* by L. Frank Baum. The small people in the story are called Munchkins.

# NEBELUNG

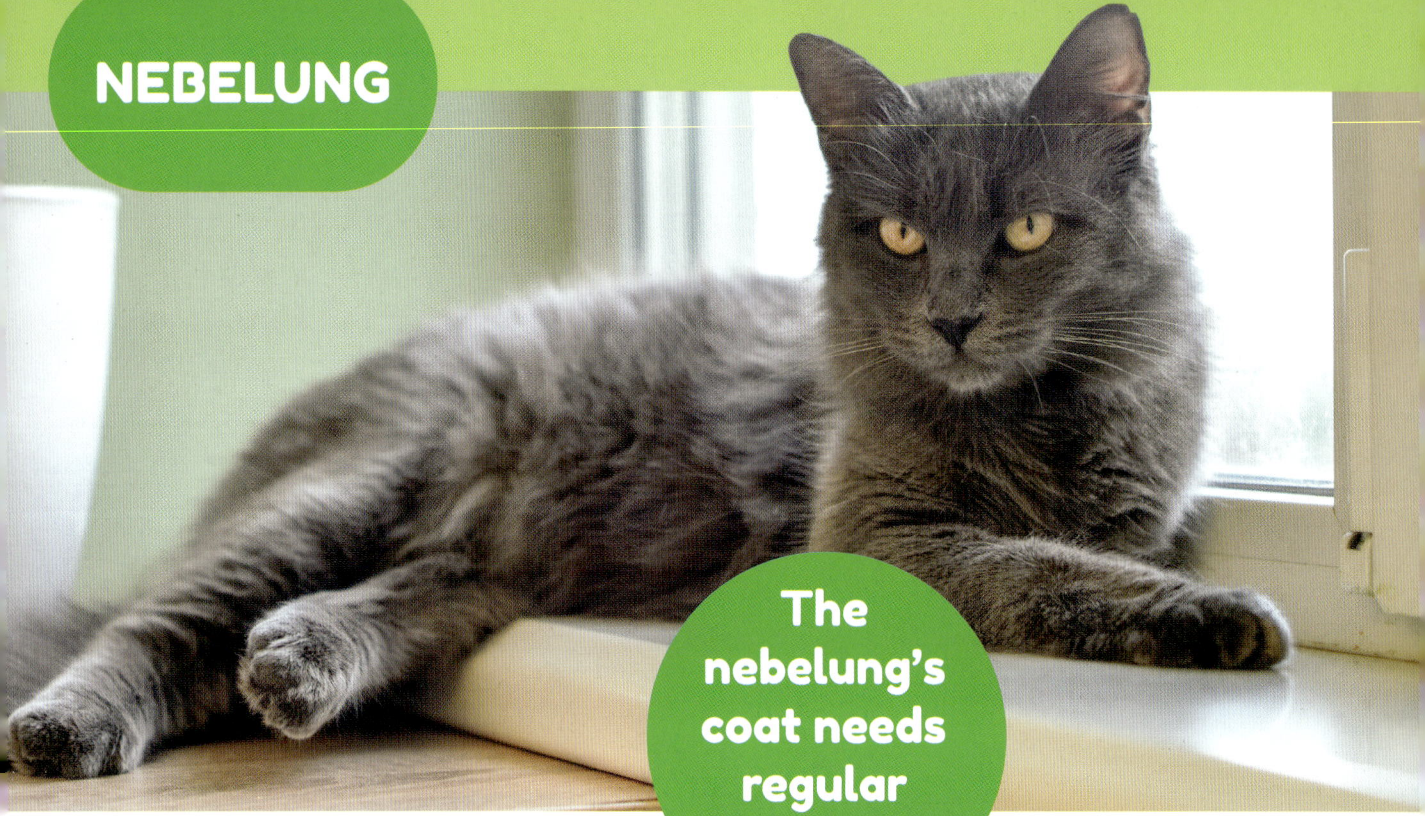

The nebelung's coat needs regular combing.

## Appearance

Nebelung (NAY-bell-ung) cats have soft, thick, blue-gray fur. It may be silver at the tips. It is longer on the legs and tails. These cats have large, pointed ears. Their eyes are green or yellow green.

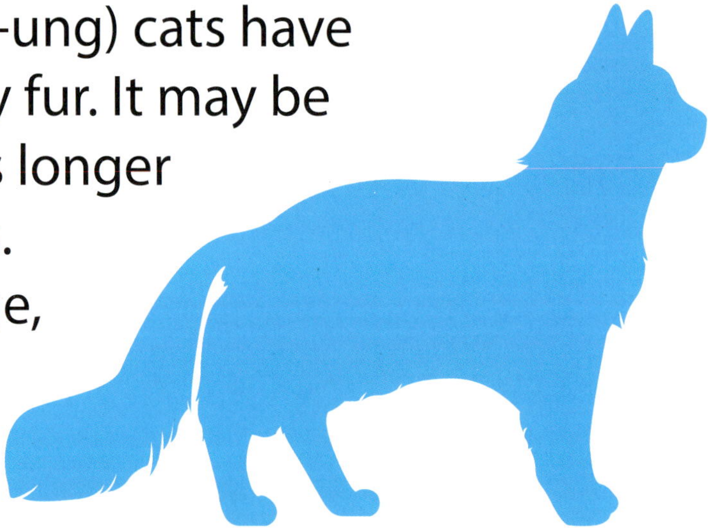

7 to 15 pounds (3.2 to 7 kg)

## Behavior

These cats love to cuddle with their people. But they will hide if someone new comes in. They do not like change. They like quiet. When they feel safe, they are playful. They like to jump and climb.

## History

In 1986, two longhaired cats with blue-gray fur had kittens. Some of the kittens had the same color. They were the first nebelungs. This name comes from the German word *Nebel*, meaning "fog."

**From:** The United States of America

## Appearance

The Norwegian forest cat has long fur. The undercoat is thick. The top coat is long and smooth. These cats shed a lot. They have long fur around their necks called ruffs. Their furry ears may be tufted. Their paws are tufted too.

**Weight:**
8 to 18 pounds
(3.6 to 8 kg)

## Behavior

These cats are kind and gentle. They have soft voices and are always ready to play. Norwegian forest cats love water. They are also good climbers.

## History

This breed comes from Norway. It is cold there, so the cats grew thick fur to stay warm. They have been around for at least 500 years. They sailed on Viking ships.

**Norwegian forest cats like to be near people.**

**From:** Norway

# OCICAT

## Appearance

Ocicats have spots like wild cats. They have long bodies. Their long tails have dark tips. The cats' short fur comes in 12 different colors. Their large ears may be tufted.

**Weight:**
8 to 12 pounds (3.6 to 5 kg)

## Behavior

These cats move like wild cats. They walk low to the ground. Ocicats jump and climb. They love people. They do not want to miss any fun. These cats are smart. They can open doors.

## History

In 1964, a breeder in Michigan mixed Siamese, Abyssinian, and American shorthair cats. One of the kittens had spots. He looked like a jungle cat called an ocelot. So the breeder called the new breed ocicat.

**Some ocicats play fetch.**

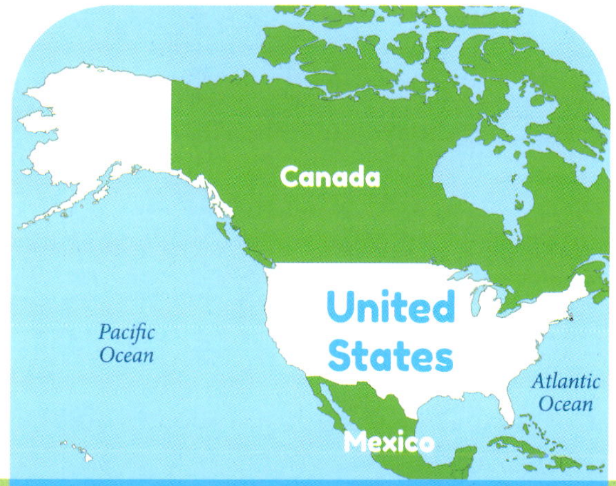


**From:** The United States of America

# ORIENTAL

Orientals come in more than 150 different colors and patterns.

## Appearance

Orientals have long, strong bodies. Their fur may be short or long. These cats have large ears and long noses. Their legs and tails are long and thin.

## Behavior

These cats need to be around people. At a party, they will visit every lap.

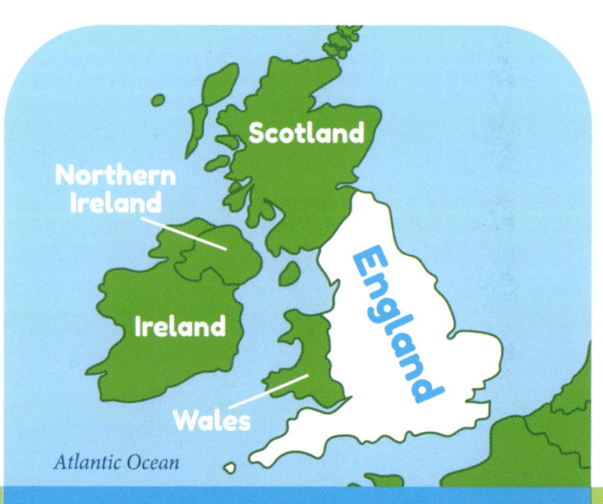


**From:** England

They are noisy. They meow, trill, chirp, and yowl. They are smart and curious. Orientals may open drawers to explore. They like to climb.

## History

Starting in the 1950s, breeders in England crossed Siamese with other breeds. The new cats had long bodies like Siamese. But their fur had many different colors and patterns. Because of this, the Oriental is also called the rainbow cat.

**Weight:**
5 to 12 pounds (2.3 to 5 kg)

# PERSIAN

Persians with nose wrinkles need their faces cleaned often.

## Appearance

Persians have long, thick fur. It comes in many colors. These cats have large, round heads. Their noses are very short. Their small ears tilt forward. They have flat faces and fat cheeks. Their eyes are large. Persians have short, fluffy tails.

**Weight:**
7 to 12 pounds (3.2 to 5 kg)

## Behavior

These cats are friendly and sweet. They will sit quietly on a chair or windowsill. They do not like surprises. They do not mind being alone sometimes. But they like people and other pets.

## History

This breed comes from Iran. This area used to be called Persia. Ancient writings from more than 3,000 years ago show cats like Persians. Persians are the most popular cat breed in the United States.

**From:** Iran

# PETERBALD

Peterbalds are fast.

## Appearance

Peterbalds can be hairless. Or they may have a bit of fuzzy or wiry fur. Some have regular fur. It can be any color. Their whiskers are crinkly. Their heads, bodies, legs, and tails are long. They have large, pointed ears. Peterbalds also have long, webbed toes.

**Weight:**
5 to 10 pounds
(2.3 to 5 kg)

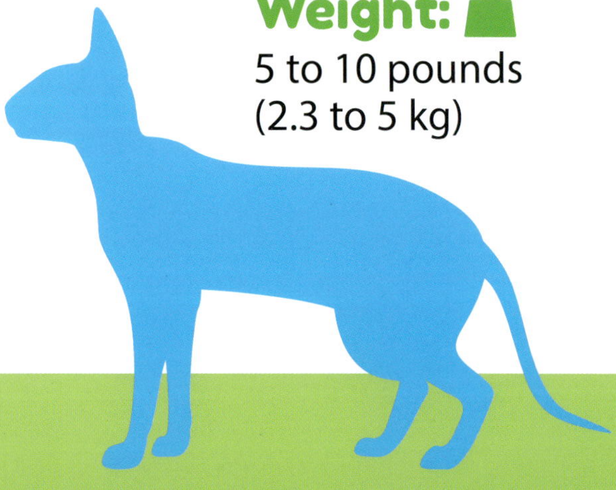

## Behavior

These cats are very friendly. They stick close to their people. They will meow to talk to their owners. They get along well with other pets. These cats hold toys with their webbed toes. They eat a lot of food.

## History

The first Peterbald was born in 1994. It was from Russia. A breeder crossed a shorthaired Oriental and a Don sphynx.

**From:** Russia

# PIXIEBOB

Pixiebobs want to be part of the family's activities.

## Appearance

Pixiebobs have short tails. Their ears may be tufted. They have brown spotted tabby coats. Their fur is woolly.

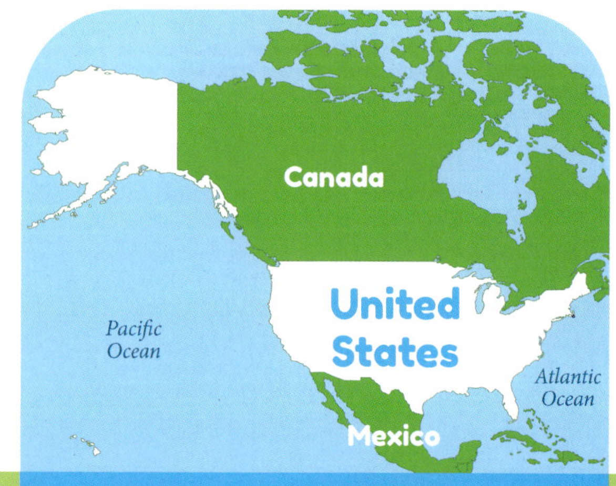


**From:** The United States of America

It may be short or long. Many Pixiebobs have extra toes. These cats are large.

**Weight:**
8 to 17 pounds (3.6 to 8 kg)

## Behavior

These cats are friendly and calm. They are smart. They like to play fetch. They like to be with their people. They love water.

## History

A woman in Washington State found a cat with a short tail. The cat had kittens with a house cat. One of the kittens looked like a bobcat. Her name was Pixie. She was the first Pixiebob. The breed was named after her.

Female RagaMuffins are much smaller than males.

# Appearance

RagaMuffins have large, long bodies. They can be up to 33 inches (84 cm) long from nose to rump. They have round heads and full cheeks. Their silky fur is medium or long. It comes in many colors and patterns.

**Weight:**
8 to 20 pounds (3.6 to 9 kg)

## Behavior

These cats are sweet and gentle. They love to be held. They will go limp in a person's arms. They love tummy rubs.

## History

In the 1960s, the Ragdoll breed started. It could be only a few colors. Other breeders wanted different colors. So they mixed Ragdolls with other longhaired breeds. They called the new breed RagaMuffins.

**From:** The United States of America

## RAGDOLL

Ragdolls are usually gentle and quiet.

## Appearance

Ragdolls have long, silky fur. They have pointed markings that can be six different colors. Those colors are seal, blue, chocolate, lilac, red, and cream. Ragdolls have large, blue eyes.

## Behavior

These cats like to play. But they like cuddling more. They love to sit on laps.

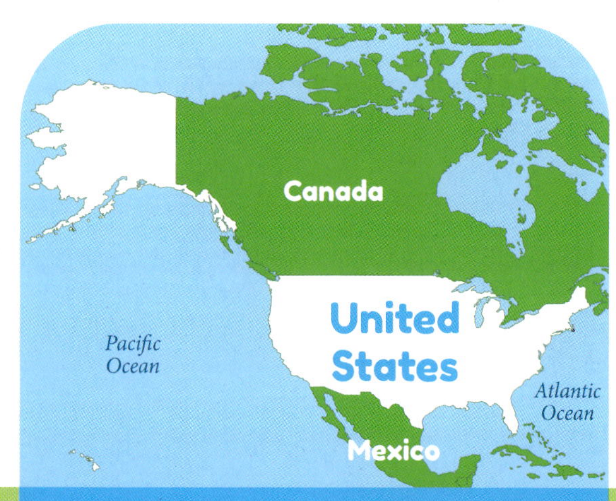


**From:** The United States of America

They want to be close to people. They get along well with other pets.

**Weight:**
10 to 20 pounds (5 to 9 kg)

## History

In the 1960s, a breeder in California crossed a longhaired cat and a Burmese-like cat. Their kittens were gentle and sweet. The breeder bred more cats with pointed markings. She called them Ragdoll cats because they would sit in a person's lap like a ragdoll. This type of doll is usually made of cloth.

# RUSSIAN BLUE

## Appearance

Russian blues have short, thick fur. The fur is blue gray with silver tips. Russian blues have wide-set, green eyes. Their bodies and legs are long and strong. They may be 24 inches (61 cm) long from nose to rump.

**Russian blues shed less than some other cats.**

**From:** Russia

## Behavior

These cats are shy with people they do not know. But they love their own people. They sense moods. If someone is sad, a Russian blue may act silly to cheer them up. They like to jump and climb. They will perch in high places.

## History

These cats came from a port city in Russia. They traveled to other cities on ships. Russian rulers may have kept these cats as pets too.

**Weight:**

7 to 12 pounds (3.2 to 5 kg)

## SAVANNAH

Savannahs' legs are long and strong.

## Appearance

Savannahs look like wild cats. Their short fur can be black, brown spotted tabby, black-silver spotted tabby, or black smoke. Savannahs have big ears. Their necks and bodies are long and lean. They can be up to 22 inches (56 cm) long from nose to rump.

**Weight:**
12 to 25 pounds (5 to 11 kg)

## Behavior

These cats like water. They are smart and curious. They love to play, jump, and climb. These cats do best with older children.

## History

A kitten named Savannah was born in 1986. Her mother was a house cat. Her father was an African wild cat called a serval. Savannah had spots and was gentle. The new breed was named after her.

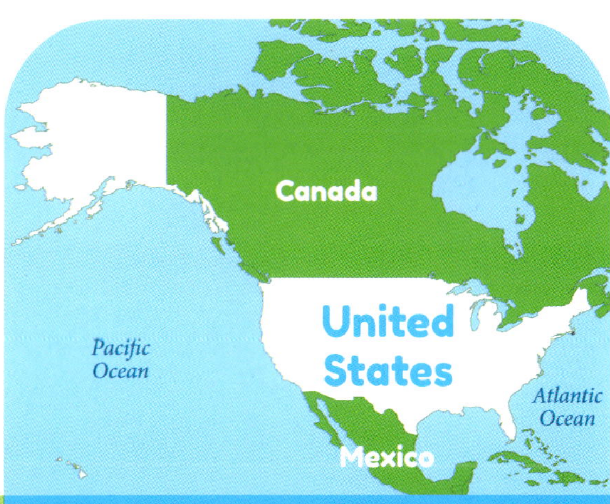


**From:** The United States of America

# SCOTTISH FOLD

Scottish folds have soft voices.

## Appearance

Scottish folds have small ears that fold forward. They have short necks and fat cheeks. They have big, round eyes and short noses. They can have short or long fur. Their fur comes in many colors and patterns.

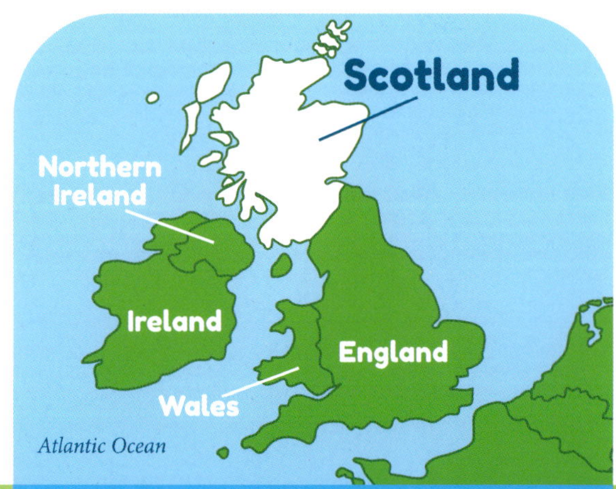


**From:** Scotland

## Behavior

These cats are sweet and quiet. Sometimes they sit on their rumps. Then they rest their front paws on their bellies. Scottish folds do not mind other pets. They do not like to be alone.

**Weight:**
7 to 10 pounds (3.2 to 5 kg)

## History

In 1961, a barn cat was born in Scotland. She had folded ears. Her name was Susie. Later, some of her kittens also had folded ears. All Scottish folds are related to Susie.

# SELKIRK REX

Longhaired Selkirk rexes need more combing than shorthairs.

## Appearance

The Selkirk rex is known for its soft, curly fur. But some have straight fur. This breed's fur can be long or short. The fur can be many colors and patterns. Selkirk rexes have curly whiskers too. Their heads are round. They have large eyes.

**Weight:**
6 to 16 pounds (2.7 to 7 kg)

## Behavior

These cats are very loving. They need to be around people. They like children. Sometimes they act silly. But this breed is also calm.

## History

In 1987, a cat with curly fur was born in a Montana animal shelter. A breeder took the kitten home. Later, some of this cat's kittens had curly fur too. The breeder named the new breed Selkirk rex after her stepfather.

**From:** The United States of America

## SIAMESE

Siamese have thin bodies.

# Appearance

Siamese have short fur with pointed markings. They have large ears. Their oval-shaped eyes are bright blue. Their eyes slant up toward their ears. Their bodies, necks, legs, and tails are long.

**From:** Thailand

## Behavior

These cats meow a lot to chat with people. They want to be in the middle of things. They will sit on laps and sleep in people's beds. They get upset if their people do not play with them. They like to jump and climb.

## History

This is one of the oldest cat breeds. They came from Siam (now Thailand). A book from about 900 years ago shows these cats.

**Weight:**
5 to 12 pounds
(2.3 to 5 kg)

# SIBERIAN

## Appearance

Siberian cats are big and strong. They are 17 to 25 inches (43 to 64 cm) long from nose to rump. Their long fur has three layers. It keeps water out. It gets thicker in the winter. It comes in all colors and patterns. The Siberian has long fur around its neck. This is called a ruff.

**From:** Russia

## Behavior

Siberians love to learn. These cats will do tricks. They like to jump and balance on things. They sense when people are sad. They will stay close. They mew, trill, chirp, and purr.

**Weight:**
Up to 25 pounds (11 kg)

## History

These cats come from Siberia in Russia. It is very cold there. Their thick fur kept them warm. They have been around for at least 1,000 years.

**A Siberian's fur needs frequent combing.**

# SINGAPURA

A Singapura's fur is easy to care for.

## Appearance

The Singapura is a very small cat breed. Singapuras are only 9 to 12 inches (23 to 30 cm) long from nose to rump. They have large ears and eyes. They have very short, silky fur. It is a color called sepia agouti (uh-GOO-tee). The base color is light. Over that is a brown ticked tabby pattern.

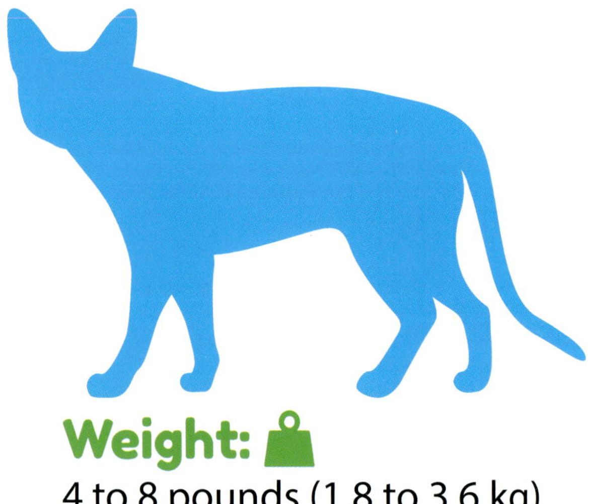

**Weight:**
4 to 8 pounds (1.8 to 3.6 kg)

## Behavior

These cats are nosy. They want to be in the middle of things. They have lots of energy. They explore and climb. When they rest, they like to be very close to people.

## History

These cats came from the streets of Singapore. Singapura is the Malaysian word for Singapore. People started breeding these cats in the 1970s.

**From:** Singapore

# SNOWSHOE

Snowshoes are easy to train.

## Appearance

Snowshoes get their name from their four white paws. They have short coats that are pointed brown or blue.

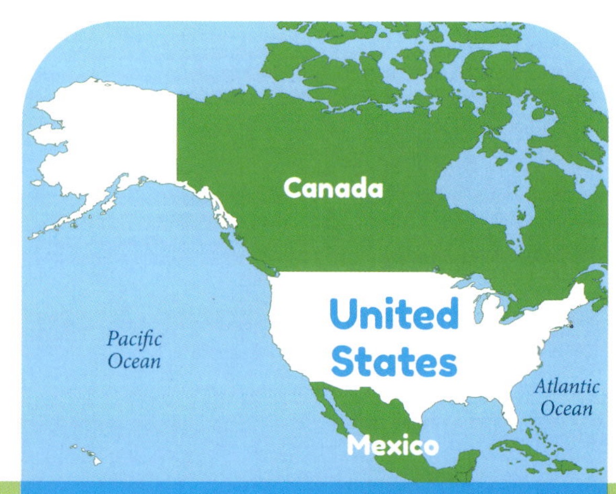


**From:** The United States of America

Snowshoe kittens are born all white. As they get older, their color comes in. They have white patches over their noses. They have large, blue eyes.

## Behavior

These cats are active. They can run fast. They like water. They are shy with strangers. But they like to be with one special person. They meow a lot. But they have soft voices.

## History

In the 1960s, a breeder in Pennsylvania found three Siamese cats with white paws. She bred them with American shorthairs to make snowshoes.

**Weight:**
7 to 12 pounds (3.2 to 5 kg)

## SOKOKE

The Sokoke's eyes can be amber to light green.

### Appearance

The Sokoke (so-KO-key) has tabby markings. But its fur is ticked all over. The fur is short and glossy. These cats have long bodies and legs. They have large ears.

### Behavior

These cats like to meet new people. They sense feelings. If someone has a bad day, they will try to cheer that person up.

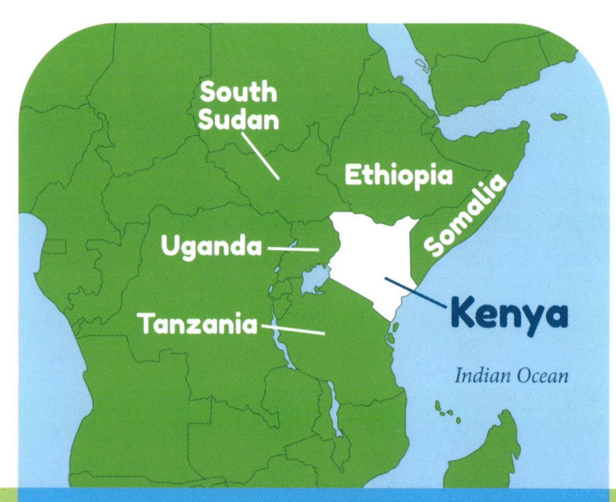


**From:** Kenya

These cats are smart. They use their meows to talk to their people. They are always ready to play. They are good jumpers.

## History

These cats come from the Arabuko Sokoke Forest in Kenya. The people there call them Kadzonzo. People started breeding these cats in the 1970s.

**Weight:**
8 to 14 pounds (3.6 to 6 kg)

## SOMALI

## Appearance

Somalis have long, soft fur. It is ticked. It comes in shades of red, blue, and fawn. Somalis have long, bushy tails. The fur on their tails and hind legs is dark. These cats have large, pointed ears. Their eyes may be gold or green.

**Weight:**
8 to 12 pounds
(3.6 to 5 kg)

## Behavior

These cats are playful. They love to climb and explore. They play with people's hair. Somalis like to do whatever their people are doing. They are smart. They can open doors.

**Somali kittens are born with blue eyes.**

## History

These cats are like Abyssinians, but they have long hair. Their name was chosen because Somalia is near Abyssinia. The breed may have started in India.

**From:** India, possibly

## SPHYNX

# Appearance

Sphynx cats look hairless. But they have some fur on their feet, noses, ears, and tails. Their skin is warm and soft. It can be many different colors and patterns. Most sphynx have no whiskers.

# Behavior

These cats love to curl up with people to stay warm. They have lots of energy.

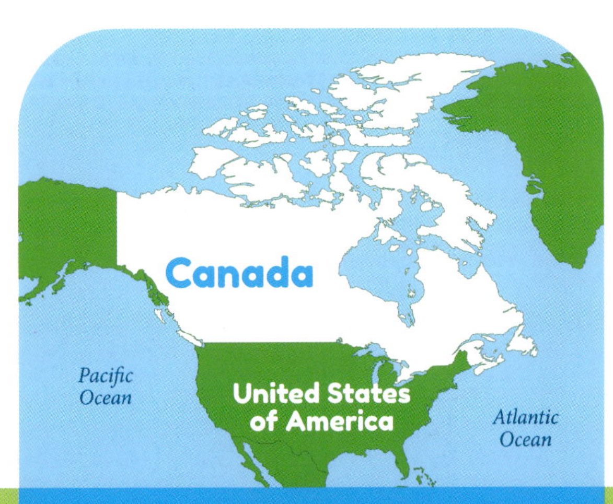


**From:** Canada

They will show off. They meow to talk to their people. They do not like to be alone. They eat a lot.

**Weight:**
6 to 12 pounds (2.7 to 5 kg)

## History

In 1966, a hairless cat was born in Canada. Later, she had some kittens that were hairless too. The new breed looked like the Sphinx, a famous Egyptian statue. So the new cats were called sphynx.

**A sphynx's skin can have wrinkles.**

# THAI

**Weight:**
7 to 12 pounds
(3.2 to 5 kg)

## Appearance

Thai cats have short fur with pointed markings. They have blue eyes. Their bodies and tails are long. They have large ears. Their heads are wide.

Thais are very smart.

## Behavior

These cats love people. To show they want to play, they might tap a person with a paw. They make many sounds to talk to people.

## History

These cats are from Thailand. A book from about 600 years ago has a poem about them. In the 1800s, British cat breeders found these cats. They called them Siamese. They bred them to make their bodies and heads longer. In the 1950s, cats with the old look became a separate breed called Thai.

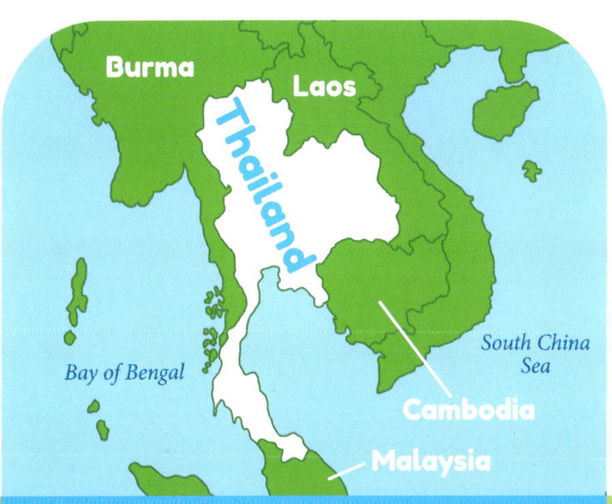


**From:** Thailand

# TONKINESE

## Appearance

Tonkinese have short, soft, silky fur. The fur comes in 12 different colors and patterns. All have darker pointed markings. These cats have almond-shaped eyes. Their eyes may be aqua, blue, green, or yellow green.

**From:** Asia

## Behavior

A Tonkinese is like two cats in one. One cat plays hard. That cat climbs, runs, and acts silly. The second cat is cuddly and loving. That cat likes to give kisses. Tonkinese can be stubborn. But they are also sweet.

**Weight:**
6 to 12 pounds (2.7 to 5 kg)

## History

This breed is from Asia. The name comes from Tonkin, the ancient name of Vietnam. Tonkinese are a cross between Siamese and Burmese cats. Some people call them chocolate Siamese.

**Owners sometimes shorten the Tonkinese name to just Tonk.**

## Appearance

Toybobs are named for their size and tails. The cats are small. Their tails are short like a bobcat's tail. Toybobs' fur may be short or long. It can be any color or pattern. These cats have large eyes.

**Weight:**
3 to 5 pounds (1.4 to 2.3 kg)

## Behavior

These cats are playful all their lives. They love to climb. They also like to sit on people's laps. They will pick one person to be their best friend. They will guard that person.

## History

In the 1980s, a Russian woman took in two stray cats with short tails. They had a kitten named Kutciy. He was very small. His tail was short too. Kutciy was the first toybob.

**From:** Russia

# TOYGER

**Toygers can learn to walk on a leash.**

## Appearance

Toygers look like small tigers. They have long, strong bodies. They have short, thick fur. The fur is striped like a tiger's. The circular markings on their faces look like a tiger's too. Their long tails have black tips.

**Weight:**
7 to 15 pounds (3.2 to 7 kg)

## Behavior

These cats look wild. But they are not. They are quiet. They love people. They get along well with other pets. They will play fetch.

## History

In the 1980s, a cat breeder wanted to make a cat that looked like a tiger. She crossed a striped shorthaired cat and a Bengal. Cat breeders are still working to make the toyger look more like a tiger.

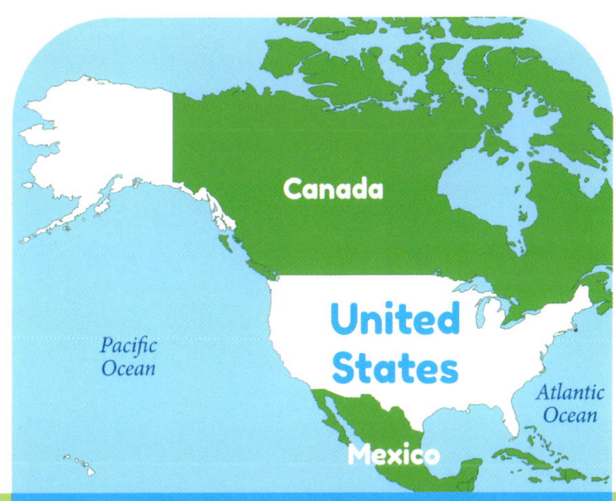


**From:** The United States of America

# TURKISH ANGORA

At first, only white Turkish Angoras could be shown at cat shows in the United States.

## Appearance

Turkish Angoras have long, shiny fur. The most common color is white. But their fur can be many colors and patterns. They can have one blue eye and one eye of another color. They have long, graceful necks.

Their small paws have tufts between the toes. They have long, brushy tails.

**From:** Turkey

## Behavior

These cats love to show off. They run, play, and climb. They are good swimmers too. They are smart. They can follow commands. Turkish Angoras like people. At a party, they will greet everyone.

## History

These cats come from Ankara, Turkey. Angora is the old name for Ankara. These cats may have been the first longhaired cats in the world.

**Weight:**
5 to 10 pounds (2.3 to 5 kg)

# TURKISH VAN

Turkish vans have strong hind legs.

## Appearance

The Turkish van (VON) has mostly white fur. These cats have darker fur on their heads and tails. Their fur is somewhat long. It keeps out water. Their eyes can be amber, blue, or two different colors. They have pink noses.

**Weight:**
7 to 20 pounds (3.2 to 9 kg)

## Behavior

These cats are called the swimming cats. They love to play in water. They also love to run. They jump with their powerful legs. They are curious. They can be a handful. But they are loving. They will follow their people around.

## History

These cats come from eastern Turkey. Many towns there have the word *van* in their names. That gave the breed its name.

**From:** Turkey

# GLOSSARY

**breeder**
A person who pairs animals together to make offspring.

**fawn**
A light grayish-brown color.

**grizzled**
Streaked with gray.

**lilac**
A shade of gray that looks light purple.

**marbled**
Having blotchy markings that look like marble.

**pointed**
Having markings on the ears, paws, tail, and face that are darker than the body coat color.

**sable**
A rich brown that is a bit lighter on the underside of the cat.

**seal**
A dark brown color.

**shaded**
A fur pattern in which the top quarter of each hair has a darker color.

**smoke**
A fur pattern in which the top half of each hair has a darker color.

**stocky**
Thick in build.

**tabby**
A fur pattern in which darker swirls, stripes, or spots of solid fur are mixed with paler ticked fur.

**ticked**
A tabby pattern in which each hair has bands of different colors.

**tufted**
Having a small cluster of fluffy hair sticking up from it.

**undercoat**
Short fur that is often partly covered by another layer of longer fur.

# TO LEARN MORE

## More Books to Read

Jenner, Caryn. *Cats and Kittens*. DK, 2020.

Mills, Andrea. *The Everything Book of Cats and Kittens*. DK, 2018.

Murray, Julie. *Cats*. Abdo, 2019.

## Online Resources

To learn more about cats, please visit **abdobooklinks.com** or scan this QR code. These links are routinely monitored and updated to provide the most current information available.

# PHOTO CREDITS

Cover Photos: Shutterstock, snowshoe; Kirill Vorobyev/Shutterstock, front (Bengal); Nynke van Holten/Shutterstock, front (Persian); Prasert Wongchindawest/Shutterstock, front, (floor); Eric Isselee/Shutterstock, back

Interior Photos: Nynke van Holten/Shutterstock, 1, 32 (top), 39 (right), 41 (bottom), 49 (bottom), 55 (top), 58 (top), 59 (top), 63 (bottom), 69 (top), 79 (top), 81 (right), 86 (top), 101 (top), 104 (top); Johanna Mehrke Fotografie/Shutterstock, 4; Eric Isselee/Shutterstock, 5, 11 (right), 19 (top), 21 (right), 23 (top), 35 (top), 38 (top), 51 (top), 64 (top), 65 (bottom), 85 (right), 98 (top), 99 (right), 117 (bottom); Oleksandr Volchanskyi/Shutterstock, 6 (top), 12, 13 (top), 26, 44, 48 (top), 66 (top), 67 (bottom), 78 (top), 110 (top); Anna Zakharchenko/Shutterstock, 6 (bottom); Shutterstock, 3, 7 (top), 8, 9 (bottom left), 10 (bottom), 13 (bottom right), 14 (bottom), 17 (top), 18 (top), 18 (bottom), 20 (top), 20 (bottom), 23 (bottom right), 24 (top), 25 (top), 27 (bottom right), 29 (top), 30, 31 (top), 31 (bottom right), 33 (top), 35 (bottom), 36 (top), 36 (bottom), 37 (top), 39 (left), 41 (top), 42 (top), 43 (bottom), 45 (top), 45 (bottom right), 47 (left), 47 (right), 49 (top), 50 (bottom), 52 (bottom), 54 (top), 55 (bottom), 56 (bottom), 59 (bottom), 60, 61 (middle), 63 (top), 65 (top), 67 (top), 68 (bottom), 73 (top), 71 (top), 74 (top), 74 (bottom), 75 (top), 76 (bottom), 78 (bottom), 81 (left), 82 (bottom), 84 (bottom), 87 (top), 88 (bottom), 89 (top), 91 (top), 93 (left), 94 (bottom), 96 (top), 97 (top), 97 (bottom), 98 (bottom), 100 (top), 101 (bottom), 102 (top), 103 (top), 103 (bottom), 104 (bottom), 106 (top), 107 (bottom), 109 (bottom), 110 (bottom), 111 (top), 112 (top), 113 (top), 114 (left), 114 (right), 117 (top), 118 (top), 118 (bottom), 119 (top), 120 (bottom), 122 (top), 123 (top), 123 (bottom), 124 (bottom); Ievgeniia Miroshnichenko/Shutterstock, 9 (top); Linn Currie/Shutterstock, 10 (top), 52 (top), 62 (top), 76 (top), 91 (bottom), 116 (top), 124 (top); iStockphoto, 14 (top), 15 (right), 40 (top); Tracy Morgan/Dorling Kindersley ltd/Alamy, 16; Dorling Kindersley ltd/Alamy, 17 (bottom right); Borkin Vadim/Shutterstock, 22; Andrey Kuzmichev/Shutterstock, 25 (bottom); Oksana Susoeva/Shutterstock, 27 (top), 43 (top); Zhu Yufang/iStockphoto, 28 (top); Irina Riedel/Shutterstock, 29 (bottom); J. E. Jevgenija/Shutterstock, 33 (bottom); Michael Hahn/Shutterstock, 34 (top); M. Davidova/Shutterstock, 46 (top), 72 (top), 73 (bottom); Ardea.com/Jean-Michel Labat/Mary Evans Picture Library/SuperStock, 50 (top), 70 (top), 71 (bottom); Perry Harmon/Shutterstock, 53 (top); Den Kara/Shutterstock, 56 (top); Arthti Kholoet/iStockphoto, 57 (right); Mila Photos/Shutterstock, 61 (top); Dorling Kindersley Universal Images Group/Newscom, 68 (top); Absolut Images/Shutterstock, 77 (right); Irina Nedikova/Shutterstock, 80 (top); Petr Jilek/Shutterstock, 82 (top); Irina Oxilixo Danilova/Shutterstock, 83 (top); Trapeza Studio/Shutterstock, 84 (top); Yves Lanceau/NHPA/Photoshot/Newscom, 87 (bottom); Tatiana Makotra/Shutterstock, 88 (top); Marina Zharinova/Shutterstock, 90 (top); Kirill Vorobyev/Shutterstock, 92 (top); Anna Utekhina/Shutterstock, 93 (right); Kolomenskaya Kseniya/Shutterstock, 94 (top), 95 (right); Krissi Lundgren/Shutterstock, 105 (top); Chris Brignell/Shutterstock, 107 (top); Minden Pictures/SuperStock, 108 (top); Omer Farukguler/Shutterstock, 109 (top); Andrey Kuzmin/Shutterstock, 113 (bottom); Benjamaphon Duangpanta/Shutterstock, 115 (top); Danilo Biancalana/Shutterstock, 120 (top); Kutikova Ekaterina/Shutterstock, 121 (top); Lea Rae/Shutterstock, 125 (top)

All maps created by Red Line Editorial.